An Acquisition Strategy, Process, and Organization for Innovative Systems

John Birkler, Giles Smith, Glenn A. Kent, Robert V. Johnson

National Defense Research Institute | RAND

Prepared for the
Office of the Secretary of Defense

The research described in this report was sponsored by the Office of the Secretary of Defense (OSD). The research was conducted in RAND's National Defense Research Institute, a federally funded research and development center supported by the OSD, the Joint Staff, the unified commands, and the defense agencies under Contract DASW01-95-C-0059.

ISBN: 0-8330-2802-2

Published 2000 by RAND
1700 Main Street, P.O. Box 2138, Santa Monica, CA 90407-2138
1333 H St., N.W., Washington, D.C. 20005-4707
RAND URL: http://www.rand.org/
To order RAND documents or to obtain additional information,
contact Distribution Services: Telephone: (310) 451-7002;
Fax: (310) 451-6915; Internet: order@rand.org

There is a widespread belief that in the future some military threats might be quite different from those of the recent past—requiring a response by innovative system and employment concepts. Our present weapon-system acquisition process was designed for a different environment than the one that exists today and seems ill suited to meet demands posed by the apparent expansion of unconventional and asymmetric threats. In response to this new environment, recent calls for reform have placed special emphasis on the need for an acquisition process that is better able to satisfy the need for truly innovative system concepts.

Most critics of the weapon-system acquisition process have emphasized the need for better performance, but rarely do they address the question of how to achieve such improvements. In this study, we examine some specific dimensions of improvement that are needed and suggest a broad strategy for achieving those improvements. The report should be of interest to those critics mentioned above and all those interested in the weapon-system acquisition process in general.

We outline a suggested acquisition strategy, process, and organization that would operate *in conjunction with* the present process, and that is specially designed for effective development of novel and more risky system concepts. We describe this strategy in enough detail to identify its key elements, suggesting how it could work and why we believe it would provide needed capabilities not present in our current acquisition process.

Some of the concepts and strategies introduced here depart substantially from current practice and would be challenging to implement. Without minimizing such difficulties, we believe the general concepts suggested herein deserve consideration in the ongoing search to improve the process for acquiring new weapon systems.

The research for this report was conducted for the Director, Acquisition Program Integration in the Office of the Under Secretary of Defense for Acquisition and Technology and was performed in the Acquisition and Technology Policy Center of RAND's National Defense Research Institute, a federally funded research and development center supported by the Office of the Secretary of Defense, the Joint Staff, the unified commands, and the defense agencies.

CONTENTS

FIGURES

SUMMARY

Our current force-modernization strategy and associated weapon-system acquisition procedures were developed in a relatively stable era of known threats and supported a steady and systematic upgrading of a large force structure. Those strategies and procedures were forged over several decades and reflect the accumulated experience of that period. But the era we are now entering is dramatically different from the recent past. The demands of the future will differ from the traditional demands of the cold war era in three important respects:

1. *Budgets will be smaller.* This puts new constraints on the traditional major defense programs that involve highly refined combat systems built in relatively large numbers and operated over long periods of time.

2. *Response times will be shortened.* In the future, the composition of potential belligerent forces and their weapons are likely to be varied, and some can be expected to appear with relatively short notice. This puts new demands on the timeliness of response by the acquisition process when called upon to deliver new kinds of systems to counter those new challenges.

3. *Novel system concepts and employment concepts will be needed.* The new forces challenging the United States are likely to include at least some that operate in very unconventional ways and that require response mechanisms not found in the established U.S. force structure.

The characteristics of novel systems are very different from those of the systems for which the present acquisition process was designed. They are so different that we believe "tinkering" with the present process will be an inadequate solution. We need an *additional* process that will be much more effective in the timely exploitation of new and novel concepts in order to be fully responsive to emerging needs.

What do we mean by "novel" systems? The main dimensions by which novel systems are expected to differ from legacy systems are shown in Figure S.1. We believe that this class of systems has several important attributes:

- The design of the system is sufficiently new, in overall concept, or in use of emerging technologies, or both, so that the development outcomes (mainly capability and cost) cannot be confidently predicted on the basis of studies alone.

- The operational employment has not been clearly defined and demonstrated and is therefore subject to substantial uncertainties and change as experience is accumulated.

RAND*MR1098-S.1*

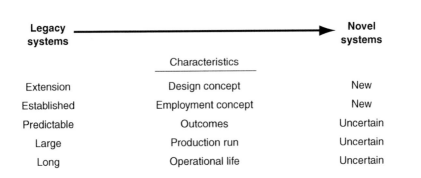

Figure S.1—Novel Systems Have Different Characteristics

- The eventual size of the weapon-system production run, and the subsequent operational life, are uncertain (an obvious consequence of the uncertainties surrounding the cost, capabilities, and operational concept of the system).

- The key uncertainties are, by their very nature, resolvable only through development and test of a demonstrator or prototype, at a cost that is commensurate with the potential value of the system.

TODAY'S ACQUISITION PROCESS IS NOT RESPONSIVE TO THE NEW CHALLENGES

It is not surprising that today's process is too narrowly focused and is not responsive to these new challenges. In particular, the standard weapon-system acquisition policy and processes are simply too risk-averse to enable the effective development and employment of new system concepts that involve some combination of true urgency and considerable uncertainties.[1] The average cycle time (program start, at milestone I, to initial operational capability, or IOC) for major acquisition programs conducted over the past several decades has averaged between eight and nine years. Some efforts last far longer. Such processes are clearly unable to meet an important segment of future demands.

Simply tinkering with the present process will not provide adequate response to future needs. The reluctance to develop bold and innovative concepts is rooted in the risk aversion that is deeply imbedded throughout the process. New, innovative concepts inherently pose many uncertainties of development outcomes (cost and performance of the system) and uncertainties of operational effectiveness. Today's process virtually demands that major uncertainties be resolved before starting major system development, thus essentially denying the start of novel concepts, or at least demanding a long, careful program of demonstration and risk reduction before starting development of the weapon system itself.

---

[1] Evidence of this exists in the well-known practice of managing such developments outside the standard process.

A NEW PROCESS FOR MODERNIZING OUR MILITARY CAPABILITIES IS NEEDED

To surmount these problems, we suggest a new strategy that should enable weapon-system acquisition managers to be more responsive to the needs of the new era. We envision a process that would operate in conjunction with the present process, which would be retained for managing large, evolutionary systems. The strategy we outline would supplement today's process by providing three capabilities that are now largely absent:

- Foster new concepts for systems and new concepts of operations.

- Enable accelerated development and demonstration of new concepts at the subsystem and system levels.

- Provide for early, "provisional" operational employment of new systems well before completion of all engineering refinements and the full set of operational testing.

The first element of this strategy is needed to provide a rich menu of new options in a timely manner. The elements of such a menu should be constrained only by perceptions of technical feasibility and operational value—not by current doctrine or force composition and employment.

The second element is needed to accommodate the kind of truly novel system and employment concepts that we will need to fully counter emerging threats and to fully exploit fast-moving new technologies. Developing such systems on a reasonably compressed schedule involves taking risks that are inconsistent with today's weapon-system acquisition policy and process. Managers must operate in an environment that views an occasional unsuccessful project as an acceptable price for meeting new operational needs.

The third element of the strategy is needed to enable the earliest possible operational employment of a new system that has been successfully demonstrated. When justified by a pressing operational need, new systems could be quickly introduced to special operational units, and the experience thus gained could help define needed design refinements.

Each of these elements represents a radical, but necessary, departure from the basic risk-averse strategy now employed for major defense acquisition programs (MDAPs). Each element has been applied in the past under special circumstances, frequently with beneficial results. But such "special-case" application hinders the systematic accumulation of management expertise and process. Hence we believe it appropriate and desirable to devise a formal, "standard" path for acquisition of novel systems, based on the strategy elements listed above. Such a path is envisioned as being *in addition to* the present standard path that would continue to be used for more-traditional programs.

A NEW CONCEPT-DEMONSTRATION PATH SHOULD BE CREATED

The main elements of the proposed new process and how they are linked together are shown in Figure S.2. The process starts with an explicit function called "concept formulation" that provides a menu of new opportunities. This menu is created by a combination of conceivers and decisionmakers. Conceivers begin the work of formulating concepts for accomplishing military tasks by thinking broadly— ideally, with no constraints other than the laws of physics. After further consideration of the technical and operational potential of various possible approaches, some of the concepts formulated will appear to be worthy of more detailed definition in terms of particular existing or emerging technologies and specific operational tactics. Consideration of the existing state of the art and the potential for advances will result in a still smaller number of concepts that appear sufficiently promising to justify the expense of actual proof-of-principle demonstrations. The primary product of this first step of the force-modernization process is a list of concepts that are technically feasible, operationally relevant, and ripe for further development.[2]

---

[2]Of course, the process needs to be iterative. Often strategic planners specify needs when a technical path gives some hope of satisfying those needs. Ideas and information must percolate upward to the strategic planner as well, so that needs can make use of options enabled by new technologies.

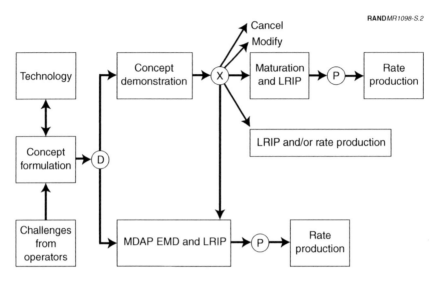

NOTES: LRIP = low-rate initial production.
        EMD = engineering and manufacturing development.

Figure S.2—Overview of Our Recommended Process

Probably the most critical and different-from-today aspect of the proposed process is that at the initial milestone review, the review authority will have *two* options for advancing a proposed new project. If the project represents a reasonably mature system concept and employment concept, with few apparent risks or uncertainties, and the anticipated production run is relatively large, then the project would be forwarded to the standard acquisition process as defined in the Department of Defense directive 5000 series policy documents. However, under the proposed process the milestone review authority will also have an additional option. If the new system concept is novel in important ways, thus presenting important unknowns and risks, it can be forwarded to the concept-demonstration process. This option is intended to be an opportunity to start such projects and to perform limited development on a speculative basis, without demanding extensive documentation purporting to demonstrate the unique correctness and sufficiency of the proposal.

By adding a new concept-demonstration path that can accommodate greater risk, it should be possible to introduce a broader menu of new concepts. By performing early development and demonstration of the critical elements in those new concepts, subsequent full acquisition and procurement of those deemed to be needed could occur much more rapidly and with greater confidence in outcomes.

Starting a concept-demonstration project carries no authority or obligation for any action beyond that phase. Activities are focused on designing and building a few items suitable for demonstrating the technical and operational capabilities of the weapon system, and performing the necessary tests and demonstrations.

At the end of concept demonstration, there is a second milestone review (milestone X). Here several options are available to decision-makers. The program might be

1. canceled

2. modified and possibly restarted as another concept demonstration

3. forwarded directly to further maturation efforts and low-rate initial production

4. passed directly into low-rate production and delivery to operating forces

5. inserted into the standard engineering and manufacturing-development (EMD) process for further refinement, maturation, and eventual large-scale production.

For programs following the concept-demonstration path, large-scale commitments and funding obligations would not be made until milestone X. At that point, far more would be known about the system capabilities and costs than typically exists at milestone II today. Furthermore, any subsequent developments needed should be completed in less time than a full EMD program would require because of the knowledge gained in the concept-demonstration design and test activities.

A New Organization Is Needed to Manage the Concept-Demonstration Path

Fundamental changes in acquisition strategy and process will change the nature of tasks to be managed and how the Office of the Under Secretary of Defense for Acquisition and Technology should be organized to perform these functions. We believe that an organization containing the following functional responsibilities would be appropriate. First, a **Science and Technology Office** would have responsibility for identifying new technologies and seeing that selected ones mature, especially those identified in the Concept Formulation and Development Office. Second, a new **Concept Formulation and Development Office** would have the charter to formulate, evaluate, and define concepts in each mission area. For selected concepts offering significant improvements in capability, the office would do the detailed end-to-end planning, take into account problems of engineering and support—e.g., joint command and logistics support—and provide oversight management to the ongoing concept-demonstration programs. Finally, an **Acquisition Office** would oversee the final acquisition of platforms and systems.

Implementation

Major changes in any large government bureaucracy are exceedingly hard to effect, and proposals for such change must be approached with respect for those difficulties. We identify some important problems that would be expected if the proposed system were to be implemented, and we believe that those problems can be resolved. Without minimizing such difficulties, we believe the general concepts suggested herein deserve consideration in the ongoing search to improve the process for acquiring new weapon systems.

Chapter One

INTRODUCTION

The length of time required to move a weapon system through the full sequence of events in the acquisition cycle has long been a source of concern and frustration to government and industry officials responsible for equipping our armed forces. The notion of somehow shortening the cycle duration has been a recurring theme in studies of acquisition and Department of Defense (DoD) management performed by various panels and commissions, and the subject of periodic initiatives by various officials.[1]

Today these desires for reform are taking on a new urgency and a new focus because of the many changes occurring in the nation's defense needs. In contrast to the relative stability and perceived predictability of the cold war years, we expect the future to involve a more rapid rate of change in both the demand and supply sides of military preparedness.

On the demand side, the traditional threats have by no means vanished, and major force elements must be retained (and improved) to counter those threats, but there is a broad consensus that a somewhat different and varied set of challenges is emerging. In a recent

---

[1]An excellent survey of such efforts is contained in *Defense Acquisition: Major U.S. Commission Reports (1949–1988)*, Washington, D.C.: U.S. Government Printing Office, prepared for the Defense Policy Panel and the Acquisition Policy Panel of the Committee on Armed Services, House of Representatives, November 1, 1988.

Naval War College publication, Admiral J. Paul Reason described the situation:[2]

> It would be dangerously imprudent for the military and political leaders of the United States to think that because American armed forces appear to be stronger than any others, they are also smarter than others are and have no critical vulnerabilities. Such arrogant opinion can become fatal delusion, for there are many asymmetric threats, and more are coming. Low-tech, self-sacrificial, asymmetric, unconventional (including but not limited to chemical and biological weapons)—these adjectives describe the kinds of threats the U.S. forces are unaccustomed to countering. Dangerous already in themselves, such threats are actually more sinister because they are stealthy: They do not appear on late–Industrial Age mental or institutional radar screens.

Similar changes are needed on the supply side, as Richard Hundley observed in a recent RAND report:[3]

> The history of the 20th century has clearly shown that advances in technology can bring about dramatic changes in military operations, often termed "revolutions in military affairs" or RMAs. Technology-driven RMAs have been occurring since the dawn of history, they will continue to occur in the future, and they will continue to bestow a military advantage on the first nation to develop and use them. Accordingly, it is important to the continued vitality and robustness of the U.S. defense posture for the DoD R&D community to be aware of technology developments that could revolutionize military operations in the future, and for the U.S. military services to be on the outlook for revolutionary ways in which to employ those technologies in warfare.

To meet these challenges, we believe that two new kinds of capabilities will be needed:

---

[2]Admiral J. Paul Reason, *Sailing New Seas,* Newport, R.I.: Naval War College, Newport Paper #13, 1998.

[3]Richard O. Hundley, *Past Revolutions, Future Transformations: What Can the History of Revolutions in Military Affairs Tell Us About Transforming the U.S. Military?* Santa Monica, Calif.: RAND, MR-1029-DARPA, 1999.

- New *operational* capabilities—to meet the different and varied challenges to our forces.

- New *institutional* capabilities—to rapidly equip our forces with new operational capabilities.

In this report, we make no attempt to predict the specific kinds of operational capabilities that might be needed in the future. In fact, we believe that one of the dominant features of such capabilities is their relative unpredictability, compared with the needs of the past several decades. Here we focus on the processes needed to enable the national defense agencies to respond to those emerging needs in a timely and effective manner.

A NEW PROCESS IS NEEDED FOR MODERNIZING FORCE CAPABILITIES

The current weapon-system acquisition strategy, process, and organizational structure is not well suited to satisfying the new and dynamic needs of the warfighters. Three aspects of the new environment are especially important to the design of a force-modernization strategy and process:

1. *Budgets will be smaller.* Development and procurement costs continue to increase for the most recent versions of traditional legacy systems, while the budgets for such acquisitions have been reduced. Compared with the peaks of the mid-1980s, the military research, development, test, and evaluation (RDT&E) budgets have been reduced by roughly one-third and the procurement budgets by nearly two-thirds. This puts special constraints on the traditional major defense programs that involve highly refined combat systems built in relatively large numbers. It seems unrealistic to expect that such large acquisition programs will continue to represent a majority of acquisition activities in the future.

2. *Response times must be shortened.* While the cold war presented very challenging threats to U.S. forces, those threats tended to evolve reasonably predictably and over relatively large periods of time. In the future, potential belligerent forces and weapons that might challenge the United States are likely to be much more

varied, and some can be expected to appear on relatively short notice. This puts new demands on the timeliness of response by the modernization process when called upon to provide new kinds of capabilities to counter those new challenges.

3. *A greater emphasis will be placed on novel concepts of systems and on new concepts of employing existing and emerging systems.* The new challenges to the United States are likely to include at least some forces that operate in very unconventional ways, and that require responses not found in the established U.S. force structure. This means that our force-modernization process must enable us to *create new and novel capabilities quickly*, with somewhat less emphasis than in the past on maximizing the cost-effectiveness of each particular system.

These demands are very different from those over the past several decades, and it is not surprising that today's process is not as responsive to the new demands as it should be. In particular, the standard process is simply too cumbersome, risk-averse, and slow to permit effective management of new concepts that involve some combination of true urgency and considerable risk and uncertainty. The cycle time (program start, at milestone I, to initial operational capability, or IOC) for major acquisition programs conducted over the past several decades has averaged between eight and nine years. Some efforts last far longer, achieving IOC only after 15–20 years of work. The F-22 Raptor is now scheduled to reach IOC in 2004, following over 20 years of effort. While those processes might have been tolerable during the cold war years, and still might be appropriate for large and evolutionary system concepts, they are not adequate to meet the full range of needs in today's environment.

The reluctance to develop bold and innovative concepts is rooted in the risk aversion that is deeply imbedded throughout the process. New, innovative system concepts inherently pose many uncertainties for development outcomes (cost and performance of the system) and operational effectiveness. Today's process virtually demands that major uncertainties be resolved before starting major system development. If novel system concepts are constrained to follow the same path, we either deny program start, or at least demand a long, careful program of demonstration and risk reduction before embarking on that path.

We have, of course, successfully managed the acquisition of novel weapon systems in the past. Examples such as precision-guided weapons and stealth aircraft come to mind. However, it is also true that in the vast majority of such cases, the first generation of truly novel systems was developed under ad hoc arrangements that substantially bypassed standard acquisition policy and procedures. Today we find the same situation, with advanced concept technology demonstration (ACTD) programs being employed to initiate and conduct development and operational demonstration of novel system concepts that either never would have been started or would have taken far too long to develop under standard acquisition-management procedures. Clearly, different treatment is needed for developing novel systems and for implementing new operational concepts.

STUDY OBJECTIVES AND REPORT ORGANIZATION

Calls for improving our acquisition process have been widespread for several decades. But major change in large, established institutions is not made easily or quickly, and few previous studies have been very specific in suggesting *how* to achieve the desired changes and improvements. In this study our *objective* is to describe a strategy, process, and organization that should be more flexible and fast moving in decision and execution, thus enabling a continuing process of force modernization that is more responsive to emerging needs.

We placed primary emphasis on identifying a set of policies and procedures that should be more responsive in defining, demonstrating, and developing novel and risky system concepts. We were relatively unconstrained by the present process, thus allowing the design of new processes and organizations more suited to the needs of such systems. Our goal was to suggest some fundamental strategies and to think through their implementation far enough to be assured that, if implemented, they would represent a major step toward the desired objective.

The remainder of the main body of the report describes the results of this task. Chapter Two develops the special characteristics of novel system concepts and describes the general strategy we propose for dealing with such systems. A new set of procedures for concept formulation is described in Chapter Three, and corresponding procedures for the subsequent development of such systems are outlined in Chapter Four. Chapter Five describes some key functional elements that should be included in any new organization responsible for managing such a process.

The acquisition-management process outlined in this report is different from the present process and organization in important ways, and would pose major problems to implement. In Chapter Six, we explore some of those problems and suggest some next steps that might be taken.

Strategies for meshing a novel system demonstration process with the budget process are discussed in the appendix.

Chapter Two

A STRATEGY FOR DEVELOPING NOVEL SYSTEMS AND ASSOCIATED OPERATIONAL CONCEPTS

In Chapter One of this report, we argued that force modernization in the future should put considerable emphasis on the introduction of some unconventional or "novel" system concepts. We further argued that the present acquisition process is poorly adapted to the timely definition and development of such systems and also to the implementation of new operational concepts—simply because they involve uncertainties and risks that the standard acquisition process has been deliberately designed to avoid.

In this chapter, we will first define what we mean by novel systems and describe their special features. We then describe a strategy and an associated process and organization for developing novel systems. That strategy should also be appropriate for developing more-conventional systems when there is a need for moving quickly and a willingness to take some risks to achieve a compressed schedule.

WHAT IS A NOVEL SYSTEM?

We do not attempt to identify specific system concepts as "novel." We can, however, make some useful observations about the new weapons and weapon systems we expect to be developing and fielding in the future.

The main dimensions by which novel systems are expected to differ from legacy systems are shown in Figure 2.1. The descriptive adjectives used to characterize the two ends of the spectrum are deliberately simplified to emphasize the extent of the differences. We believe that this class of systems has several important attributes:

- The design of the system is sufficiently new—in overall concept, or in use of emerging technologies, or both—so that the development outcomes (mainly performance and cost) cannot be confidently predicted on the basis of studies alone.

- The operational employment has not been clearly defined and demonstrated and is therefore subject to substantial uncertainties and change.

- The eventual size of the production run and the subsequent operational life are uncertain (an obvious consequence of the uncertainties surrounding the cost, capabilities, and operational concept of the system).

- The key uncertainties are, by their very nature, resolvable only through development and test of a system or prototype, hopefully at a cost that is commensurate with the potential value of the system.

To illustrate, consider the case of unmanned aerial vehicles (UAVs). It has been technically possible to build generic UAV platforms for

RAND*MR1098-2.1*

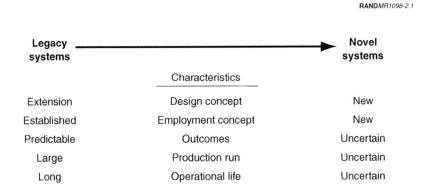

Legacy systems	→	Novel systems
	Characteristics	
Extension	Design concept	New
Established	Employment concept	New
Predictable	Outcomes	Uncertain
Large	Production run	Uncertain
Long	Operational life	Uncertain

Figure 2.1—Novel Systems Have Different Characteristics

several decades, and many have been built and used as aerial targets and reconnaissance drones. Throughout that period, various combat and combat support roles have been posited for UAVs, but every such application raised a host of troublesome issues: Exactly how would they be controlled, especially in situations demanding deviation from the original mission plan? How would the information normally obtained visually by the pilot be acquired and translated into mission-relevant decisions? How would safety be ensured during peacetime operations over populated areas? etc. Despite many studies, and a small number of actual development projects that were canceled early, few UAV programs were actually completed in the United States. How can we explain this? One attempt was made in an earlier report:[1]

> The cause of the poor track record of UAV programs in the United States is not entirely clear. Certainly, the mere fact of their being *unmanned* vehicles cannot be the cause. After all, the United States has had great success with other unmanned systems, ranging from interplanetary spacecraft and satellites to cruise missiles and submersibles. What, then, makes UAVs unique? A possible explanation is that UAVs in general have never had the degree of operational user support necessary to allow their procurement in sufficient quantities (perhaps because of funding competitions from incumbent programs, or because of the conjectural nature of their capabilities). Thus, the learning curve is never ascended, multiple failures occur, risk tolerance decreases, unit costs rise as a result, and user support decreases yet further in a diminishing spiral.

One explanation not noted in the above quotation is that the major system acquisition policy and process as defined in the Department of Defense (DoD) 5000 series of directives is not congenial to programs with a range of important uncertainties. A key problem lies in the sequence of decisions and actions involved in the process: In phase 0, a series of studies are performed, followed by a milestone I decision where the system concept to be developed is clearly defined and the sponsor commits to full funding for development and initial

---

[1] Geoffrey Sommer et al., *The Global Hawk Unmanned Aerial Vehicle Acquisition Process: A Summary of Phase 1 Experience,* Santa Monica, Calif.: RAND, MR-809-DARPA, 1997.

production.[2] In the case of novel systems with major uncertainties that are not likely to be adequately resolved through analysis, an *acquisition* program is delayed—or never gets started. Further, in the UAV case (as in all but truly exceptional cases) funding needed for a program to explore and resolve the major uncertainties exceeded that which could be obtained by a project not directly coupled with a major acquisition program. Thus, uncertainties were unresolved and progress was limited.

While not every new system developed and employed in the coming years will reflect all of the features of novel systems, we expect that an important segment of new developments will reflect many of those characteristics.

ELEMENTS OF A STRATEGY FOR FIELDING NOVEL SYSTEM CONCEPTS

The characteristics of novel systems are very different from those of the systems for which the present acquisition process was designed. They are so different that we believe "tinkering" with the present process will be an inadequate solution. To provide guidance in the formulation of new procedures, we first sought to identify a few major elements of acquisition strategy that would enable the process to deal with the special features of such systems and the expected environment of urgency that might attend their development. Five such strategy elements were selected:

- Provide an environment that fosters new concepts for systems and new concepts of operations.

- Conduct accelerated development and demonstration of new concepts at the subsystem and system level, without commitment to full procurement and fielding.

- Upon successful demonstration of a new system, permit early, provisional fielding and operation before completion of full maturation development and associated testing.

---

[2]Earlier versions of DoD directive 5000.2 made explicit the demand for full funding commitment at milestone I. That demand was dropped in the most recent revision, but the requirement is still mostly imposed in practice.

- Encourage timely and visionary decisions on such programs by enabling programs to be approved and guided by a few senior officials, without the demand for extensive staff support and documentation.

- Provide a new and separate organization to oversee the development and demonstration of novel systems and operational concepts.

The first element of this strategy is needed to provide a rich menu of new options in a timely manner. The elements of such a menu, created by a staff combining technical and operational experience and skills, should be constrained only by perceptions of technical feasibility and operational value, and not by current doctrine on force composition and employment.

The second element is needed to accommodate the kind of truly novel system and employment concepts that we will need to fully counter emerging threats and to fully exploit fast-moving commercial technologies. Developing such systems on a reasonably compressed schedule involves taking risks that are inconsistent with today's acquisition policy and process. By performing early development and demonstration of the critical elements in those new concepts, subsequent full acquisition and procurement of those deemed to be needed could occur much more rapidly and with greater confidence in outcomes.

The third element of the strategy focuses on the later phases of acquisition. Today's procedures demand extensive effort toward system maturation to minimize support costs, together with extensive operational testing to ensure that no lingering problems and deficiencies exist. This phase can be shortened by introducing early operation of the system, employing special dual-use operational units concurrent with maturation and testing activities.

The fourth element of the strategy is needed to remove the impediment of extensive and lengthy documentation and review now required for milestone approval—especially at milestone I. Those procedures were designed to ensure not only a full examination of all alternative concepts, but incidentally the creation of a broad-based consensus on the final selected concept. Such extensive reviews might make sense when the proposed new system concept is an extension of previous design concepts and operational employment strategies because of the broad accumulation of experience and historical data that are available. However, such accumulation of experience and corresponding data does not exist for novel system concepts, and extensive analytical examination and comparisons add little real value. Instead we must put much more faith in the insight, imagination, and wisdom of senior officials to make judgments and decisions, in the expectation that their decisions will turn out to be justified at least most of the time. To be effective, those officials must operate in an environment that views an occasional unsuccessful project as an acceptable price for building a menu of new projects that can be used as a base for rapidly responding to new technological opportunities and new operational needs.

The last element of the new strategy is needed because the organizations that evolved over the past several decades are so imbued with the traditional mindset that they would be ineffective in dealing with the risks and uncertainties inherent in the anticipated new system concepts. New organizations, with explicit charters for experimentation, testing and learning, and demonstration should provide a more congenial attitude for discovering and developing novel system and employment concepts and for systematically accumulating information on how to manage such a program.

Each of these represents a radical, but necessary, departure from the basic risk-averse strategy now employed for major defense acquisition programs (MDAPs). Each element has been applied in the past under special circumstances, with beneficial results. But such "special case" applications hinder the systematic accumulation of management expertise and process. Hence we believe it appropriate and desirable to devise a formal, "standard" path for acquisition of novel systems, based on the strategy elements listed above.

Readers will recognize here some features of today's ACTD process. We add one critical feature; *we integrate it into the overall force-modernization process and organization.* The ACTD process has demonstrated ability to conceive, develop, and produce small quantities of truly novel systems on an accelerated schedule, but it exists "outside the system," and the services have not exploited the opportunities it presents—especially with respect to efficiently integrating the products into the operational forces. By adopting the key aspects of the ACTD process and formalizing them as a legitimate part of the overall process, we hope to more tightly integrate the selection of such projects and their final introduction into the operational forces and into overall service force plans.

The key feature of the proposed strategy is the introduction of a second, parallel path for acquisition management: the concept-demonstration path. This path is envisioned as being in addition to the present standard path that would continue to be used for more-traditional programs.

A schematic showing the main elements of the proposed strategy is shown in Figure 2.2. The process starts with an explicit function called "concept formulation" that provides a menu of new opportunities. A project starts when a new system concept or operations concept is submitted to milestone D (development) for start of development. At this point the milestone authority[3] has two options for moving the project ahead. If the proposed system has the main attributes of a MDAP, it could be authorized to proceed along the standard management path as defined in DoD directive 5000. However, if the proposed system has the attributes of a novel system as outlined above, it could be authorized to start along the concept-demonstration path.

Starting a concept-demonstration project carries no authority or obligation for any action beyond that phase. Activities are focused on designing and building a few items suitable for demonstrating the technical and operational capabilities of the system, and performing the necessary tests and demonstrations.

At the end of demonstration there is a second milestone review (milestone X). Here several options are available to decisionmakers. The program might be (a) canceled; (b) modified and possibly restarted as another concept demonstration; (c) forwarded directly to maturation efforts (further engineering work to strengthen system integration, to improve operational suitability or manufacturing efficiency, etc.) and low-rate initial production; (d) passed directly into

---

[3]Typically, the milestone authority is the Systems Acquisition Review Council at OSD or service level.

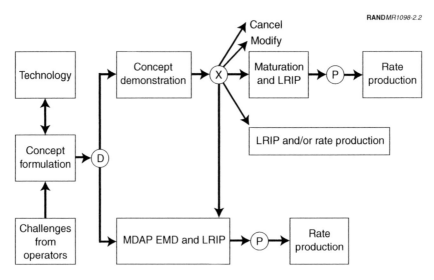

RANDMR1098-2.2

NOTES: LRIP = low-rate initial production.
        EMD = engineering and manufacturing development.

Figure 2.2—Overview of Our Recommended Process

low-rate production and delivery to operating forces; or (e) inserted into the standard engineering and manufacturing development (EMD) process for further refinement, maturation, and eventual large-scale production. The milestone review authority for milestone X is the same as that for milestone D.

Those projects that are approved at milestone X for acquisition might be transferred to the EMD phase or to the full MDAP process (but since they have already received considerable development and testing, further EMD could probably be abbreviated). Other projects might be deemed suitable for continued application of accelerated practices that would lead to earlier, but possibly "conditional," operational employment.

The proposed concept-demonstration path is built upon four key features:

- A concept-formulation phase, designed to be more congenial to novel, innovative system and operational concepts.

- A concept-demonstration phase for initial development and demonstration of new projects that do not fit well within the established weapon-system acquisition process.

- An expeditious process for transitioning from demonstration to acquisition.

- An accelerated-acquisition phase that would effectively utilize the products of the concept-demonstration phase and lead to early operational employment.

Each of these phases is described in the following chapters.

Chapter Three

CREATING AND CHOOSING OPTIONS FOR FORCE MODERNIZATION

Choosing new weapon-system concepts to be developed and implemented is always a contentious process, fraught with uncertainties. These choices are more contentious and uncertain at times like the present when international events and domestic budgets seem to require significant changes in the character and capabilities of U.S. military forces.

With major changes occurring in systems and forces and the way those forces are used, it is more important than ever that new concepts be evaluated against an integrated set of the types of operational capabilities needed to meet the needs of the future. The reasons for modernizing certain types of military forces in the first place should be questioned anew, and the process for making modernization decisions should be examined to see if improvements in the process are warranted or possible. In this chapter, we examine these aspects of force modernization.

HOW THE PROCESS SHOULD WORK[1]

Defense planning begins with basic national-level objectives. These are found in such documents as the U.S. Constitution and are enduring and constant regardless of the geopolitical environment. Consistent with these enduring objectives, the president and his advisors set forth broad national security objectives—in the congressionally mandated *National Security Strategy of the United States* and elsewhere—toward which U.S. national power is applied. These national security objectives, which can change with international circumstances, are formulated and defined in light of U.S. interests, threats to these interests, and opportunities for advancing them.[2]

A set of national military objectives or missions is derived from these broad national-security objectives by the Chairman of the Joint Chiefs of Staff, within guidelines set by the Secretary of Defense. Broadly stated, the following are the current U.S. military missions:

- Deter and defeat attacks on the United States.

- Deter and defeat aggression against U.S. allies, friends, and global interests.

- Protect the lives of U.S. citizens in foreign locations.

- Counter regional threats involving weapons of mass destruction.

- Underwrite and foster regional stability.

- Deter and counter state-sponsored and other terrorism.

- Provide humanitarian and disaster relief to needy peoples.

Each of these missions breaks down into a number of still more specific operational objectives. For example, the following are the

---

[1]For details on how the framework relates to how the Department of Defense develops new operational concepts to enhance military capability, see Glenn A. Kent and David E. Thaler, *A New Concept for Streamlining Up-Front Planning*, Santa Monica, Calif.: RAND, MR-271-AF, 1993; Glenn A. Kent and William E. Simons, *A Framework for Enhancing Operational Capabilities*, Santa Monica, Calif.: RAND, R-4043-AF, 1991; and David E. Thaler, *Strategies to Tasks: A Framework for Linking Means and Ends*, Santa Monica, Calif.: RAND, MR-300-AF, 1993.

[2]Glenn A. Kent, *A Framework for Defense Planning*, Santa Monica, Calif.: RAND, R-3721-AF/OSD, 1989.

operational objectives derived from the military mission "underwrite and foster regional stability":

- Prevent the coercion of allied and friendly governments.
- Promote and maintain desirable regional balances of power.
- Protect threatened indigenous populations.
- Enforce the cessation of hostilities.
- Bolster democracy.

These operational objectives break down yet further into specific military tasks. For example, to "enforce the cessation of hostilities," the U.S. military has to be ready to accomplish the following tasks, among others:

- Stop or prevent artillery, mortar, and sniper attacks against designated targets or areas.
- Enforce no-fly zones.
- Resupply friendly forces and civilians.
- Neutralize enemy radars.
- Identify and disarm combatants.
- Locate and destroy weapons caches.
- Clear and avoid mines.
- Maintain persistent surveillance of selected areas.

Only when we have articulated quite specifically the nature of military tasks can we think concretely about alternative approaches to accomplishing those tasks.

The process of modernizing military forces is made up of four key steps, performed in a logical sequence by four different types of people and organizations.[3] The four steps of this process (illustrated schematically in Figure 3.1) are as follows.

_____

[3]For an in-depth discussion of the defense-planning process, of various techniques that can be useful in that process, and how concept development fits within a larger

Step 1: Strategic Planners Establish Demands for Military Capabilities

Strategic planners—typically within the staff of the Chairman of the Joint Chiefs of Staff and the Office of the Secretary of Defense (OSD)—are responsible for articulating the principal missions to be accomplished by U.S. military forces, consistent with the strategic goals set forth by high-level national authorities, including the president and his national-security advisors. Aided by operational commanders and analysts, these planners also identify the specific military tasks that must be performed if these principal missions are to be accomplished. In essence, strategic planners identify what it is that military forces have to be ready to do.

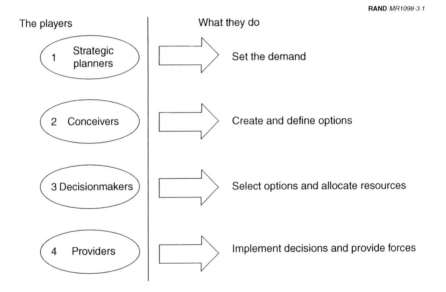

Figure 3.1—Force Modernization Involves Four Core Activities in a Logical Functional Flow

planning framework, see Paul K. Davis and Zalmay M. Khalilzad, *A Composite Approach to Air Force Planning*, Santa Monica, Calif.: RAND, MR-787-AF, 1996. This work uses the earlier name, concept *action* groups, for what we now call concept *options* groups; the activities referred to are the same.

More specifically, the planners (1) set the strategic orientation of U.S. forces and describe a vision of the operational objectives to be achieved and military tasks to be accomplished by these forces; (2) establish the characteristics of military capabilities deemed most relevant to meet future needs—the operational requirements; and (3) identify operational deficiencies that should receive priority attention.

Strategic planners provide the terms of reference for each of the subsequent steps in the process of modernizing forces.

Step 2: Creative Conceivers Formulate Options

New concepts on which to base new capabilities do not appear automatically. They arise from the concerted efforts of creative minds that understand emerging technologies, the operational realities that military commanders face, and the overall strategic environment as laid down by the strategic planners. It is, of course, rare for this breadth of understanding to be found in a single person, and consequently the process of defining systems and operational options must be pursued by teams.

Conceivers begin the work of formulating concepts for accomplishing military tasks by thinking broadly—ideally, with no constraints other than the laws of physics. After further consideration of the technical and operational potentials of various possible approaches, some of the concepts formulated will appear to be worthy of more detailed definition in terms of particular existing or emerging technologies and specific operational tactics. Consideration of the existing state of the art and the potential for advances will result in a yet smaller number of concepts that appear sufficiently promising to justify the expense of actual proof-of-principle demonstrations. The primary product of this second step of the force-modernization process is a list of concepts that are technically feasible, operationally relevant, and ripe for further development.[4]

---

[4]Of course, the process needs to be iterative. Often strategic planners specify needs when a technical path gives some hope of satisfying those needs. Ideas and information must percolate upward to the strategic planner as well, so that needs can make use of options enabled by new technologies.

Step 3: Top-Level Decisionmakers Choose Among Available Options

Top-level decisionmakers within OSD choose which options to implement. Costs are considered; implications for required manpower and training are delineated; the relevance of particular concepts to multiple missions is debated; judgments about the relative importance of various tasks and missions are made. This is the step where the difficult decisions about resource allocation are made. Top-level decisionmakers combine the demands laid down by the strategic planners and the options identified by the creative conceivers into choices about which missions and tasks to pursue, how to pursue them, and how many resources to devote to them.

Step 4: Force Providers Implement Decisions

The force-modernization process is completed when acquisition officials act on the choices made by top-level decisionmakers, undertaking and completing the development of chosen systems, procuring them, and providing them to operational commands. Force providers are also responsible for organizing, manning, equipping, training, and maintaining the force, and for developing the tactical doctrine necessary to make the chosen new concepts and associated systems operationally functional.

Structuring the force-modernization process as described above emphasizes the *functions* that must be performed in the course of this process. Having specified these functions, we can then begin to consider the roles to be played by various *organizations* in performing these functions. Relevant organizations might include various components of the Office of the Secretary of Defense, the Joint Staff, the services, various unified commands, analytic shops, think tanks, and defense contractors. Thinking first about functions and then about organizations is, we believe, superior to thinking first about organizations and then defining functions to fit the needs and wishes of interested organizations.

This vision of the force-modernization process also makes clear that it is the function of the force providers—who organize, man, equip, and sustain these force elements—to implement the decisions of the top-level decisionmakers. *It is not the role of these force providers to make decisions about resource allocations.*

Finally, and most relevant in the current context, this vision of the force-modernization process emphasizes the importance of formulating new concepts about how to accomplish important military tasks. Moreover, it underlines the fact that the objective of concept development is not to find a use for a given technology. Rather, the process of concept development begins with military requirements articulated by strategic planners and proceeds to identifying technologies that contribute to meeting these requirements. Nor is concept development concerned exclusively with hardware; formulating a concept for fulfilling a military task requires descriptions of technologies and systems *and* of how those systems are to be operated.

CURRENT APPROACHES TO FORCE MODERNIZATION ARE LESS THAN IDEAL

Unfortunately, current practice within DoD deviates from the sequence of steps described above. In particular, current practice has the effect of reversing the second and third steps of the process. (See Figure 3.2.)

RAND *MR1098-3.2*

- Develop an MNS for a particular system (looks like a ship, tank, airplane)

- JROC "validates" MNS, then the Defense Acquisition Board orders concept development

Figure 3.2—Current Practice Reverses Concept Development and Choice of a System

Current practice begins—as it should—with strategic planners articulating required military missions and tasks. But then, contrary to the logic of the force-modernization process, the military services (which are most naturally cast in the role of force providers, not creative conceivers) respond to these articulated needs with so-called mission needs statements (MNSs), which often describe particular systems that might be suitable for meeting specific needs. Only after an MNS is validated by the Joint Requirements Oversight Council (JROC) does the Defense Acquisition Board order a formal concept-development effort. In effect, an MNS documents a choice of a particular approach to accomplishing a mission and thus convolutes the process.

This reversal of steps has the effect of narrowing the range of potential concepts that receive serious consideration. In developing an MNS, a service inevitably draws on the expertise available to it, often without the benefit of expertise available in other services, in national laboratories, from contractors with which that service does not work routinely, and so on. The result is that important choices among various options are made (by default, if not otherwise) *before* a broad range of options has been formulated and considered. MNSs drafted by individual services will typically reflect the strengths of a particular service; from the outset the proposed system is likely to look like a ship, a tank, a missile, an airplane, or whatever, depending principally on which service or which part of a service is developing the MNS. By the time formal work on concept development begins, potentially interesting avenues—particularly avenues that exploit joint operations—may have been closed. The force-modernization process risks getting locked into past concepts regarding which service or which kinds of platforms will perform specific missions and tasks. Further, true competition among a broad range of alternative approaches can be precluded.

A NEW APPROACH TO CONCEIVING AND SELECTING NEW SYSTEM CONCEPTS

We believe the current process needs to be redesigned so that it performs more like the idealized process outlined in Figure 3.1 above. Two changes are needed:

- A process and organization to encourage formulation of new system or employment concepts that are innovative and boldly exploit emerging technologies and ideas.

- Providing the decisionmakers with an option for forwarding the more risky but potentially valuable concepts to a process of demonstration, without any obligation for further action until demonstration results are known.

The main elements of such a process are outlined in Figure 3.3. The conceivers continually seek new ideas for system and employment concepts, drawing on opportunities provided by the evolving technology base and demands posed by operating forces. The product of their work is a menu of system concepts that are deemed to be within the technical state of art and that, if developed, should be of value to the operational forces. That menu is not constrained by budgets, perceived priorities, doctrine, or any such considerations. It is a menu of opportunities.

RAND*MR1098-3.3*

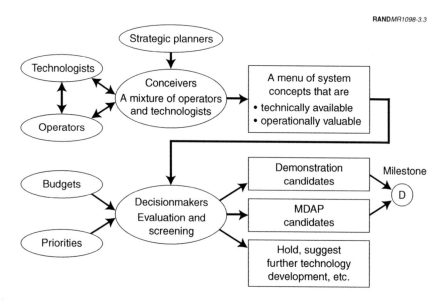

Figure 3.3—Proposed Process for Conceiving and Selecting New System Concepts

It is the task of the decisionmakers to review that menu and select from it the new system concepts that appear to be highest in priority in terms of responding to emerging operational needs, and affordable within current budget projections. Some of those options will be deemed sufficiently "mature" (low technical risk, well understood operational concept, etc.) so that they can be recommended for initiation of EMD via standard acquisition practices. However, not all chosen concepts need be that mature. Some will be sufficiently novel that they pose risks and uncertainties too large to justify immediate start of full EMD. Those system concepts can be recommended for partial development, sufficient to support testing and demonstration needed to resolve risks and uncertainties. A final decision on whether those candidates ever advance to full development and procurement will depend on the outcome of the tests and demonstrations.

We believe that if the spirit of this functional outline were followed, a more inclusive range of options would receive consideration and that the most attractive of those options would be more quickly defined and translated into concept-demonstration projects.

MILESTONE REVIEW FOR PROJECT START

As noted in Chapter One of this report, the present acquisition policy and process are not congenial to early development of novel system concepts. Those concepts inherently contain important risks and uncertainties that the present system is designed to screen out and to prevent from entering acquisition. One of the key mechanisms for such screening is the demand that the sponsoring service provide full funding for the project in the out years of the program objective memorandum (POM).[5] Thus we believe that one criterion for a new development process specifically oriented to early start of development projects for novel system concepts is to limit the objective of the initial activity to that of demonstrating the feasibility of the concept, and thus limit the commitment needed from sponsors.

---

[5]Earlier versions of DoD directive 5000.2 carried an explicit demand for such funding as a requisite to passing milestone I. That demand does not appear in the most recent revision of the policy, but it is widely believed that passage of milestone I would be difficult in the absence of such funding coverage.

The concept of a limited objective is critical. If no commitment is made beyond the demonstration of key uncertainties, then (1) more risky (but potentially valuable) projects can be started, and (2) all energy and resources can be focused on the demonstration, thus constraining the scope of activities and the resulting schedule and funding. We propose that this concept of limited objectives be explicitly stated in the policy covering this class of development activities, that several alternative outcomes of the "demonstration phase" be recognized as legitimate, and that none would be stigmatized as a "failure."

Implementation of this strategy starts with the milestone review where new development projects are approved and authorized. Probably the most critical, and different-from-today, aspect of the initial milestone review is that the review authority will have two options for advancing a proposed new project. This plan is outlined in Figure 3.4. If the project represents a reasonably mature system concept and employment concept, with few apparent risks or uncertainties, and the anticipated production run is relatively large, then the project would be forwarded to the standard acquisition process. However, under the proposed process the milestone review authority will also have an additional option. If the proposed new system concept is novel in important ways, thus presenting important unknowns and risks, it can be forwarded to the proposed new concept-demonstration process. This option is intended to be an opportunity to start such projects and to perform limited development on a speculative basis, without demanding extensive and full documentation purporting to demonstrate the unique correctness and sufficiency of the proposal.

Contrary to standards existing for typical MDAPs, we propose that minimal documentation and staffing be required to support a decision to start a concept demonstration. In addition to the description of the concept, as proposed by the conceivers and proposers or by any defense agency, the decisionmakers would typically receive three documents prepared by appropriate staff:

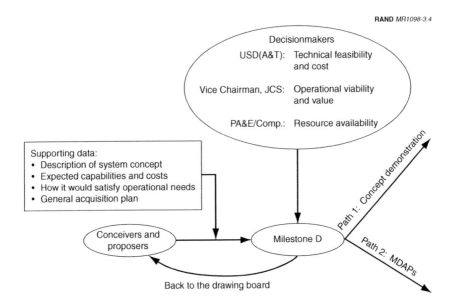

Figure 3.4—Milestone Decision Process

- An estimate of the expected costs of the proposed system.

- An assessment of the expected operational capabilities and the value of those capabilities, together with a description of a concept of operations.

- A preliminary acquisition plan, identifying the critical features of the system that are expected to be demonstrated, the test activities needed to provide the specified demonstration results, and an outline of activities that should occur during the demonstration phase to support subsequent transition to a following stage.

All such documents and plans should be tailored to the needs of the particular project and to the desires of the controlling officials. The guiding policy would be to do only that planning, and prepare only those documents, that is absolutely needed to guide designers and project managers during the demonstration phase.

The decision authority would be vested in a small group of senior officials. We suggest that approval of the following three officials, constituting a senior overview panel, should be sufficient authority to start the project:

- Vice Chairman, Joint Chiefs of Staff (JCS): attests to operational capabilities and the relevance of those capabilities.

- Program Analysis and Evaluation (PA&E)/Comptroller: attests to cost estimates and that resources are available.

- The Under Secretary of Defense for Acquisition and Technology (USD(A&T)): attests that the concept is technically feasible and the program is executable.

Each of these officials would be able to request that specific planning tasks and support documentation be provided before voting in favor of starting a project, but typically we anticipate that the list outlined above would suffice for most projects. The goal here is to enable decisions to be made quickly on ambitious and risky projects if the decisionmakers determine that they are sufficiently important to counterbalance the risk. Staff review and support is valuable on routine projects, but exceptional advances generally depend on the vision and authority of a few senior officials. The people occupying those positions of responsibility should have the express opportunity to act with corresponding authority when the opportunity arises.

Chapter Four

THE CONCEPT-DEMONSTRATION PATH

In previous chapters, we developed arguments on two features of the overall force-modernization process: first, that in the foreseeable future such modernization will depend to an increasing degree on weapons with novel design concepts and novel employment strategies and doctrines; and second, that we need to add a process that would enable the decisionmakers to authorize the start of some novel system concepts that would otherwise be considered too risky, thus creating a richer set of options available for acquisition, and that would enable efficient and timely acquisition of those systems to meet rapidly evolving operational needs.

In this chapter, we describe how the proposed concept-demonstration path should function.

THE CONCEPT-DEMONSTRATION PATH

The content of the demonstration phase must be designed to efficiently deliver the desired demonstration of system capabilities. The current ACTD process provides such an option—a process with an objective limited to that of demonstrating major uncertainties facing a particular new system or subsystem.

The overall design for the concept-demonstration phase is outlined in Figure 4.1. It consists of an initial demonstration phase followed by a special milestone review (milestone X). That review would assess the outcomes of the demonstration phase, the condition of the system in terms of maturity, and the level of operational need for the capability provided by the system.

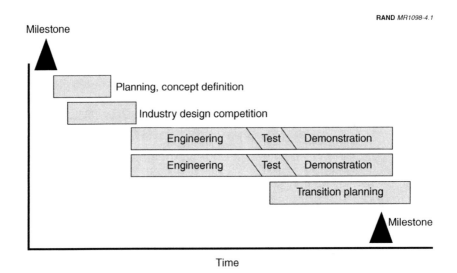

RAND *MR1098-4.1*

Figure 4.1—Main Elements of the Concept-Demonstration Path

Our vision for the design and management of the demonstration phase is guided by recent experience in ACTD programs. These include the following key features.

A limited concept-definition phase focuses on identifying and defining a general system concept and an associated operations concept. This process requires tight cooperation and coordination among technology experts and staff from service organizations responsible for developing concepts and doctrine for future forces. The product of this phase should be a system description phrased in terms of *goals* for operational capability, together with goals for costs and schedule of the concept-demonstration phase. Some sense of the relative priorities among the various goals is also desirable to provide guidance to industry when making design studies and project proposals, and to the source selection authority when choosing among candidate designs. It is especially important to provide system and project goals and constraints in the most general form possible, and to focus on desired system *capabilities* rather than system *description*, to achieve the desired economy of both time and cost for concept demonstration and to provide the broadest opportunity to industry to create innovative solutions.

The concept-definition activity should also define a general plan for the test and demonstration of the new system, and for a business strategy to be employed throughout the phase. Informal participation by industry should be employed to the extent possible to stimulate innovation and to help define realistic goals and constraints for the program. The final product should be a formal solicitation to industry for proposals on the design and demonstration of the system.

Highly streamlined management utilizes features authorized by legislation first passed for Pilot Acquisition programs and later expanded as Other Transactions Authority.[1] These authorities allow the government project office to use an "agreement" in lieu of a contract and permits the waiver of Federal Acquisition Regulations (FAR) and Defense FAR Supplements, the Armed Services Procurement Act, the Competition in Contracting Act, and the Truth in Negotiations Act, in addition to releasing the contractor from mandatory military-specification compliance. All procurement-system regulations are waived. It also frees the contractor from the need to undergo Defense Contract Auditing Agency audits, allowing instead the use of commercial auditors.

Competition continues as much as possible throughout the demonstration phase, both as a strategy to ensure the highest possible product value, and as an effective and efficient substitute for extensive management oversight. By "competition" we do *not* mean price competition, which would be completely inappropriate in this kind of activity. Instead we seek to motivate the contractor to provide the highest possible value in terms of system design and development, stimulated by the prospect of gaining a place in subsequent phases of the procurement if they are performed.

A demonstration phase designed to yield sufficient **information on system cost and capabilities,** and general validation of both design concept and operational concept, supports the decision regarding subsequent disposition of the project. If the project runs into unex-

---

[1] See Public Law 101-189, Section 2371, Title 10, USC, and Section 845 of the 1994 National Defense Authorizations Act (Public Law 103-160).

pected problems, leading to cancellation or redirection, those problems would then be understood well enough to provide a firm foundation for disposition of the project. If the demonstration was successful and the user deemed it desirable to obtain production copies for inventory, then results of the demonstration phase would provide a solid and quantitative basis for planning that subsequent work.

TRANSITION PLANNING

One of the important activities that should be conducted throughout the demonstration phase is planning for transition to subsequent phases. Such planning should be "success oriented"; that is, it should assume that the results of the demonstration phase are positive and that a decision will be made to move the program into further development and eventually production. By anticipating the needs of such subsequent work, planning can smooth the transition, reducing time and cost of the overall program. Furthermore, the planning activity itself is inexpensive, and it need not result in obligation of further work or expenditures until late in the demonstration phase when initial results are available.

The actual content of such planning would be project dependent but would typically include topics such as these:

- Drafting a plan for the following phase, including a description of tasks to be performed, a schedule of such tasks, and a funding plan that identifies the timely obligation of funds needed for expeditious progress.

- Planning for the disposition or subsequent use of the test articles fabricated for the demonstration phase and integrating such plans with related programming of user activities and facilities.

- Drafting a solicitation for quotations from the contractors for subsequent work on design engineering and preparation of production facilities and drafting related modifications to the contract or agreement needed to cover such work.

- Examining how to most efficiently sustain any tooling, facilities, and skilled staff used to fabricate the demonstration articles during any subsequent interval preceding rate production.

- Preparing an operational-support plan, defining the kinds of special support equipment needed, stocks of spare parts to be procured, etc. The extent of this activity during transition from concept demonstration to a later phase would depend on the relative maturity of the system at that point and the expectation of near-term production.

During the early portion of the demonstration phase, such planning can be focused on identifying issues and laying out a schedule of anticipated activities. As the demonstration phase progresses, information will be gained and some tentative assessment of project results can be made well before completion of final demonstration tests. If those early results are positive (suggesting technical success and continued user interest in obtaining the system), then the appropriate long-lead activities can be performed, including earmarking funds to ensure timely start of future work.

The evolution of transition planning must be tightly integrated with the user community and with the evolving outcomes of the demonstration phase. We therefore recommend that such transition planning be performed by the government project office managing the demonstration, and that the same office maintain management responsibility across the transition and throughout the subsequent phases. The issue of management continuity is discussed further in Chapter Six of this report.

TRANSITION TO ACQUISITION PHASE

One of the key features of the concept-demonstration path is that the demonstration phase can be started, and conducted, without anyone making a commitment for subsequent acquisition of the system. Thus, when demonstration has been completed, a further milestone review will be needed to validate demonstration results and authorize any further actions. We designate that as milestone X.

At milestone X, the review authority has several options as illustrated in Figure 4.2. One option is to terminate the project; it might have proved incapable of meeting goals, or it was overtaken by other projects of higher priority, etc. Existing resources (test articles, etc.) would typically be scrapped or returned for other research projects. Another option is to terminate the project but to retain the residual assets for their limited operational value. A third option is to decide the concept has promise but must be substantially revised, with a new set of capability goals so different from the original ones that substantial redesign must be accomplished. This might lead to the start of a new demonstration program. Finally, test results might be so promising as to warrant acquisition and operational employment. Exactly how that last option will be carried out is itself subject to a wide range of options, depending on the particular situation. If it is anticipated that only a small number of the design will be produced for inventory, and the design is close to adequate maturity, then a modest engineering upgrade might be undertaken, possibly in parallel with low-rate production. If a large production run is anticipated, with long-term operational employment that would justify extensive design refinement for production efficiency and improved operational suitability, the design might be passed to the established MDAP process for completion.

To achieve maximum integration of activities across the several phases of this acquisition strategy, the authority to transition from demonstration phase to maturation/LRIP (low-rate initial produc-

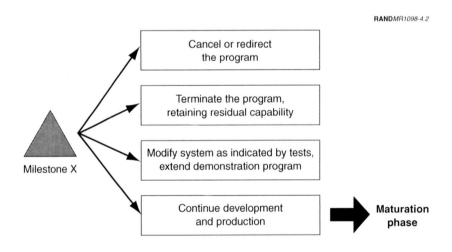

RANDMR1098-4.2

Figure 4.2—Possible Outcomes of a Demonstration Project

tion) should be the same three-person senior overview panel empowered to start the demonstration phase (see above). That panel would thus have continuous responsibility for, and control over, the total process.

Prerequisites and documentation required for the transition would be at the discretion of this panel and should be allowed to vary from one project to another depending on the type and extent of activities anticipated in the maturation and LRIP phase.

FUNDING THE TRANSITION FROM DEMONSTRATION TO ACQUISITION

The demonstration phase is typically started without any formal commitment of support beyond demonstration. If the demonstration program is successful, any attempt to quickly move the program to further development or early production and operation could be hindered because funding for such activities had not been placed in the budget during the appropriate POM cycle two to three years earlier. To achieve truly expeditious handling of such projects, some way must be found to solve the budget-cycle delay problem.

This is not a new problem in weapons acquisition management. The budget process suggests that all expenditures should be predicted at least three years before funds are needed so that those funds could be reflected in the budget as it is prepared. However, demands for funds are not always that predictable. As shifting demands place emphasis on accelerating the schedule of certain development programs, or as unexpected problems demand greater-than-scheduled funds to resolve, demands for unbudgeted funds frequently appear just before, or even during, the budget year.

One method for satisfying truly unpredictable demands for funds is to reprogram—move money from one of the budgeted accounts to satisfy the unbudgeted demand. That is almost always an inefficient process because it causes the cancellation or stretch of some other program. It might be used to provide *some* of the funds needed to support development or production of a system stemming from a successful demonstration program that was deemed worthy of further activity, but it should never be relied on as a standard policy. Thus we must look elsewhere.

One special feature of a demonstration program selected to move into further development or production is that the funds needed are not completely unpredictable; they just were not budgeted because of uncertainty over whether the funds would be needed. Thus, one option is to start budgeting such funds partway through the demonstration phase, drawing on early results (engineering tests that occur before the formal operational demonstration, or development test results, etc.). This notion is illustrated in Figure 4.3. The confidence in program outcomes tends to build to a substantial level well before completion of all tests, thus allowing introduction of some budget actions to support downstream activities well before final demonstration is complete.

As noted above, such budgeting actions should be a major task of the transition team that starts work early in the demonstration phase. This might be practical for some programs, in part because funding needed for the first year past operational demonstration might be modest while the program builds up. This approach will likely be most practical for programs in which the demonstration phase ex-

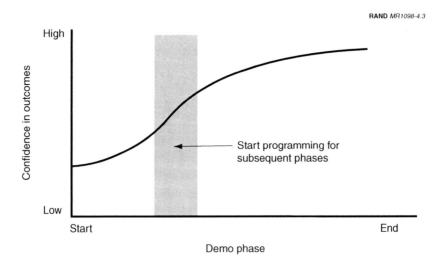

Figure 4.3—Funds Can Be Inserted in POM Before the End of the Demonstration

tends over several years. Thus, halfway through a demonstration program there might be enough information to justify inserting funds for follow-on efforts into the then-current budget process.

The need to tightly integrate the intermediate progress and expectations of a demonstration program into the budget process is another reason to seek tighter integration of such demonstration programs into the formal process for managing acquisition programs.

A third strategy exists for funding post-demonstration activities that were not incorporated in the budget during POM preparation two to three years earlier. The DoD should establish a special TBD (to be determined) fund to support such activities. Such funds would be held in a special budget category, approved by Congress for this purpose. Those funds would not be earmarked for a specific project but could be allocated by agreement of a specified group of senior officers and executives, possibly with the concurrence of a few senior members of appropriate congressional committees.

There is precedent for this approach. The Army's uniformed leadership (Chief of Staff and Commander, U.S. Army Training and Doctrine Command) in the spring of 1996 reached agreement with members of Congress to have funds (6.4 RDT&E dollars) for accelerating development and fielding of selected hardware and software. One of the key features of the initiative, called the Warfighting Rapid Acquisition Program (WRAP), was that the funds would be appropriated by Congress without the prior identification of what specific projects were to be undertaken. However, agreement was made that prior to the Army obligating any funds for a specific project, the Army would identify the projects to the four defense committees on the Hill and obtain their written approval. The level of funds appropriated was $50 million in FY97, and approximately $100 million in FY98. FY99 funding was expected to be around $70 million. Additional information on such programs is contained in the appendix.

ACCELERATED ACQUISITION

If the demonstration phase leads to a decision at milestone X that a system similar to that demonstrated is worthy of production and introduction into the operational inventory, the next step is to complete whatever additional engineering, system integration, and maturation and test activities are needed.

The content of this phase will vary over a wide range, depending on the degree of system completeness and overall design maturity achieved in the demonstrator, and on the size of the anticipated production and operations activity. At one extreme, suppose that the objectives of the demonstration phase required that the design to be demonstrated be a relatively complete system and thus much of the final system engineering and integration tasks had been completed. Suppose further that the engineering and operational tests performed during the demonstration phase had revealed no major changes needed in the design. Finally, suppose that the system concept was applied to a rather specialized and narrow operational application such that only a few units were anticipated for production and operation. Under those conditions, it should be possible to quickly perform any needed design refinements; prepare whatever provisions are needed for operational support equipment, spares, manuals, etc.; and proceed directly to a short production run.

At the other extreme, if large production quantities are envisioned, together with long-term operation of the system, considerably more development work might be needed. It might, for example, be necessary to add or modify major subsystems, make extensive design refinements to ensure reliability and supportability of the system, do major redesign to make the item production-ready, conduct a substantial test program to validate the changes, and make extensive investment in production facilities. Under those conditions, the maturation and LRIP phase could closely resemble a traditional EMD phase, and a great many of the standard EMD procedures and practices should be applicable.

Because we are dealing here with a particular class of systems that are so novel and unconventional that standard development and acquisition practices do not work very well, it seems likely that the majority of those programs will not justify EMD-like practices following concept demonstration. Thus we need to examine other practices more applicable to this class of systems.

One dominant feature that will affect the choice is that many such systems will not be planned for large production runs, at least initially. Furthermore, as information processing becomes a larger component of weapon systems, we can expect that few configurations will remain unchanged for long. Thus we can begin to examine practices specifically tailored to relatively small production runs, or large production runs that occur in groups of small quantities with major design upgrades interspersed between groups, akin to upgrades of computer software and hardware systems.

One other feature affecting design of the maturation/LRIP phase is that early delivery to the user is likely to be important. The combination of relatively short production runs together with an urgency in delivering a new capability to the user leads to a policy for the maturation and LRIP phase starkly different from that of the standard EMD phase. Instead of devoting great energies and resources to refining the design for efficient production and for a high degree of reliability and operational suitability, and then subjecting the system to rigorous testing to demonstrate that those goals had been accomplished, consider doing exactly the opposite: Take a system configuration that is "pretty good," maybe much like the one tested in the demonstration phase, and put it into low-rate production. Deliver the early production units to a special "provisional" operational unit that would contribute to a diligent process of accumulating field operating experience. Appropriate design refinements would then be engineered and incorporated into the production process as the need arises. We choose to call the initial delivery of such immature units "early operational capability" (EOC) to distinguish it from the conventional initial operational capability (IOC).

This approach might be criticized on the grounds that those initial production items are likely to harbor major flaws that will have to be subsequently corrected at great cost. Recent research indicates that is highly unlikely.[2] Major flaws in any new system (those demanding corrective action to make the system capable of performing its specified mission) are almost always detected relatively early in development testing. The sorry history of some major systems that still contained such gross problems when delivered to the user can always be traced to a lack of early action to correct known problems. Such lack

---

[2]Giles K. Smith, unpublished RAND research on use of flight test results in support of F-22 production decision.

of timely action was in turn generally driven by a perceived need to sustain production and delivery schedule at all costs. In the concept-demonstration path, we believe such major flaws would become obvious during demonstration, and the absence of a predetermined production program and funding commitment would enable managers to take appropriate corrective action.

Those major flaws detected in development testing could be scheduled for correction during maturation and before LRIP started. A completely different issue is presented by problems of system reliability and maintainability. Discovery of those typically follow a different pattern, with many not becoming apparent until some experience has been accumulated by the operational unit. Thus, a careful and systematic program of early LRIP and delivery to operating forces can be an efficient and effective way of supporting system maturation.

To make this process work, some accommodation must be made by the system operators in terms of both unit staffing and measures of unit performance. Instead of the usual policy of staffing such a unit with a normal cross section of skills and experience levels, an EOC unit should include some personnel with special technical skills who can support the immature system and the engineering design team in identifying configuration refinements. Likewise, the operational command must give formal recognition of such activity by modifying its measures of unit performance and providing suitable incentives for managing the design refinement process as well as training for operational employment of the unit.

This process is not without precedent. The F-117 Stealth Fighter followed almost exactly the EOC process because in that program considerable emphasis was placed on achieving the earliest possible operational capability. That emphasis, combined with the many special design problems imposed by the stealth performance goal, led the designers to defer many maintainability refinements until after the initial lot supporting development testing was produced. The low production rate of eight per year that was sustained for that airplane made it eminently practical to follow this process, and supportability refinements were introduced over the first two to three years after delivery of the initial units to the operating command.[3]

Obviously, the application of this tactic must be tailored to the particular needs of each system. When fully applied, we believe that duration of the concept-demonstration phase plus the maturation/LRIP phase, measured until EOC, could be 20–30 percent shorter than the same system developed under conventional procedures.

System Testing

In any acquisition program, testing plays an important role. In the traditional process, the testing performed during EMD has two major objectives. First, engineering tests are conducted to ensure that the system is safe to operate and performs according to the specifications laid down at the beginning of EMD. Second, operational testing is conducted to ensure that the system performs suitably in a realistic operational environment. Typically, an OSD-level director of operational test must observe the operational tests and certify to Congress that the system meets the operational performance and capability goals established in the test plan before the Congress will authorize expenditure of funds for high-rate production. This can cause a considerable problem of scheduling because the operational tests must be conducted with early production articles that are representative of the expected high-rate production items. But if a year or two passes after that initial production, while operational tests are under way, a major disruption is created in the production process. That problem has been ameliorated by the introduction of low-rate initial production, whereby the production process is continued during operational testing but at a low rate.

Under the accelerated-acquisition plan, we propose a substantially different approach to the testing phase. Instead of focusing testing on the demonstration of a complete and mature system ready for high-rate production, the initial period of testing should focus on

---

[3]See Giles K. Smith, Hyman L. Shulman, and Robert S. Leonard, *Applications of F-117 Acquisition Strategy to Other Programs in the New Acquisition Environment,* Santa Monica, Calif.: RAND, MR-749-AF, 1996.

those aspects of the system deemed necessary for initial, "provisional" operation, while accepting some limitations in both functional performance and operational suitability.

During the demonstration phase, the contractor should be required to demonstrate general safety of operation and the ability to achieve the main functional performance goals sufficiently to justify start of system-capability-demonstration testing. Throughout the demonstration phase the system contractor(s) should be closely involved so that they can monitor system performance and identify design modifications that should be addressed during maturation and LRIP.

To further accelerate the program, testing at this stage should *not* be required to demonstrate functional performance throughout the entire operational envelope, only that portion near the design points where the system is expected to be operated most of the time. Nor should it be necessary to demonstrate full compliance with reliability and maintainability (R&M) goals. The justification for this limited set of maturation test objectives is the observation that if a system is going to experience major failures or problems sufficient to bring the entire project into question, it is highly likely that those problems will be revealed during the earliest part of the test program. If a system passes through the initial testing during both the demonstration phase and the maturation phase, it is highly likely that any remaining flaws will emerge only through extensive operation of the system and that those future flaws will be fixable with reasonable expenditure of time and money.

The early production items will be delivered to a provisional operational unit that is charged with two objectives. First, to provide a limited but useful combat capability in case the system is called upon to support some contingency; and second, to work with the contractor and system manager in extending the demonstration of both functional and R&M performance. By combining early operational capability with such extended testing, it should be possible to significantly shorten the time to achieve that limited operational capability, while routine training operations can partially support test data collection that would otherwise have to be charged entirely to development costs. The detailed planning and organization of this combined operations and test activity will depend on the individual characteristics of each system, but adoption of this general strategy should save both time and money.

Production Phase

As in the previous phases, the production of a novel system is likely to benefit from a somewhat nonstandard policy and treatment, stemming from two broad characteristics that are expected to be typical: first, it is likely that the system will be modified and upgraded frequently during its operational life, drawing on rapid advances in information processing and other technologies. Second, the production runs are not expected to be large, since most such systems are expected to fill niches (albeit very important niches) in the broader force structure.

In traditional production programs of major weapon systems, production rate is generally strongly influenced by economic considerations: how rapidly funds can be made available to the program; the rate at which the operational forces can absorb the new design, with the attendant burdens of training, provisioning, etc.; and the desire to minimize overhead charges by clumping the total production in as few years as possible. These influences have tended to yield relatively high-rate, short-duration production runs. Yet the system characteristics of the special class of novel systems examined here seem to argue for relatively low-rate, extended-duration production runs. Such a production profile should make possible the incorporation of block upgrades before completion of production and fine-tuning of total planned quantities based on extended feedback from operational experience.

One immediate criticism of such a policy is likely to focus on expected cost increases relative to a higher-rate plan. However, much of the folklore supporting such an expectation is based on programs that were planned, facilitized, and staffed for one rate, and then that rate was reduced. A number of reviews have documented such cost increases and developed rudimentary models for predicting their scope.[4] However, that is not the model we suggest here. Instead, we propose a general strategy of deliberately planning, facilitizing, and staffing for a production rate somewhat lower than would be traditionally applied. If properly done, this should permit a direct production unit cost not much greater than that stemming from a higher rate, and even those costs might be offset by more efficient introduction of upgrades and redesigns to improve operability. Several prior studies suggest the practicality of such a plan. For example, the F-117 was produced at a rate of eight per year with no apparent cost penalty associated with such a low rate.[5] Similarly, an earlier study of the A-7D suggested that an extended period of low-rate production lasting some two years beyond EMD could have yielded cost savings due to more efficient incorporation of design changes to fix avionics performance and reliability problems.[6] These examples strongly suggest that an extended period of relatively low-rate production need not incur an economic penalty if properly planned and executed.

A lower-than-typical production rate could be expected to yield positive benefits for many of the systems considered in this report by providing an opportunity to introduce modifications and upgrades quickly and efficiently as technology advances and as operational needs evolve. Commercial industries have made great strides in the past decade in reducing supply pipeline times, thereby increasing their ability to shift production from one model to another. While shifting production to a modified or upgraded version is somewhat different from the commercial industry model, many of the same techniques should be applicable to timely incorporation of such modifications or upgrades in a military system.

Finally, for those systems that seem likely to follow the pattern outlined here, with a relatively small total production run that incorporates even smaller groups of models with different technology levels and different mixes of mission equipment, the use of contractor lo-

---

[4]See *Effects of Weapons Procurement Stretch-Outs on Costs and Schedules,* Congressional Budget Office study, November 1987. Also, Steve J. Balut, "Redistributing Fixed Overhead Costs," *Concepts,* Vol. 4, No. 2, Spring 1981.

[5]RAND, MR-749-AF.

[6]J. R. Nelson et al., *A Weapon-System Life-Cycle Overview: The A-7D Experience,* Santa Monica, Calif.: RAND, R-1452-PR, 1974.

gistic support should be considered for at least the initial years of system operation. The prime system contractor is likely to be much better prepared to support the evolving configuration than is the established service depot system.

We therefore envision the overall production phase of novel systems as incorporating several features:

- Rate production would be authorized at some point during maturation and LRIP when technical and operational experience with the LRIP models indicate sufficient system maturity.

- The subsequent "rate" production might not be much greater than during LRIP.

- Planning and engineering for modifications and upgrades should be planned as part of the production phase, thus facilitating smooth incorporation of such versions into the ongoing production.

- Contractor logistic support is likely to be appropriate, for at least the initial years of system operation.

Some will argue that what we are proposing is the equivalent of "concurrent" development and production, a process much criticized in prior programs because it yielded some early production items being delivered to the field in an immature configuration, requiring subsequent and costly retrofit. In many instances, that criticism appeared justified for three reasons. First, production rate was accelerated early, leading to a relatively large number of immature products that needed retrofit. In our plan, we recognize the probability that those early items will need corrective work, so we keep the early production rate low, thus allowing time to learn about the problems and take corrective action before many items are produced. Furthermore, since that activity is anticipated, resources (time and money) will have been earmarked for it.

A second major source of problems with earlier "concurrent" programs was that the early items were delivered to an operational unit that was not staffed, trained, or prepared in any way to cope with the problems arising from the immature design. In our plan, we propose that the operational unit receiving the initial novel-system units be specifically prepared to support the continuing design and maturation activity through appropriate record-keeping, and that the op-erational unit would be supported by initial contractor logistic sup-port during the transition period.

Finally, in the earlier programs that were criticized because of prob-lems stemming from concurrency, little or no credit was taken for early operational capability. In at least some cases, such early capa-bility should be accorded some value to partially offset any addi-tional costs incurred.

Nonetheless, this process will need to be specifically tailored to the special features of each program.

ACCELERATING THE PROCESS

One of the main objectives sought in the design of this proposed new process was to accelerate the schedule for moving new system con-cepts to the operational user. How do we expect such acceleration to occur?

Acceleration of the development and acquisition cycle should occur in four ways. First, an explicit and continuing activity of concept formulation should generate new ideas and opportunities more rapidly; second, by enabling selected concepts to start actual devel-opment even though important risks and uncertainties are involved, it should be possible to start such meaningful efforts several years ahead of the otherwise-typical start date. *We substitute action for analysis.* There will be some "losses" associated with such a plan, but that is an inevitable and acceptable consequence of achieving rapid-response capability.

Third, the early start of new projects will be enhanced through a streamlined milestone review process, with limited documentation required. In some cases, the review authority might demand exten-sive analysis and documentation, but in other cases the review au-thority is expected to rely on good judgment, supported by less for-mal and less fully staffed analysis.

Finally, if the concept-demonstration phase yields a system or sub-system that appears to satisfy an immediate need, and the design being demonstrated appears sufficiently mature to justify early op-erational employment, then some limited operational capability can be achieved up to several years before the system passes through the "normal" regimen of development testing, operational testing, and design refinement.

This proposed process does not provide a magic trick that would dramatically shorten the acquisition cycle duration for every new system. It does provide the flexibility to define new opportunities and to enable some system concepts to be developed relatively quickly if the requisite sense of urgency seems justified.

Chapter Five

AN ORGANIZATION TO MANAGE MODERNIZATION

Fundamental changes in the strategy and process for modernizing our forces will change the nature of tasks to be managed and how the office of USD(A&T) should be organized to perform these functions. In this chapter, we identify some functional relationships that should be provided within the office of USD(A&T) in order to properly implement and manage the strategy and process outlined in previous chapters. No attempt is made here to define a complete organization for the office of USD(A&T) or to show how these special functions fit into the overall organization; such implementation details go far beyond the scope of this study.

ORGANIZATION OF STRATEGIES

We begin by identifying the five principal activities of USD(A&T):

- Discovering new technologies.

- Demonstrating selected technology aggregates and subsystems.

- Participating in formulating and defining new system concepts and new operational concepts.

- Managing the acquisition of systems and weapons.

- Advising the Secretary of Defense about matters of modernizing.

A set of functions and relationships that accommodates these activities is shown in Figure 5.1.

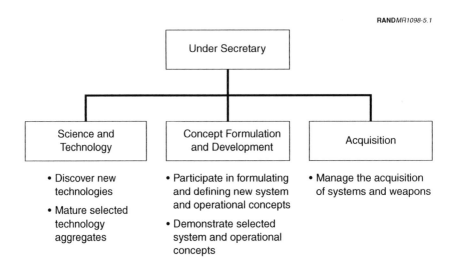

Figure 5.1—A Suggested Functional Organizational Structure

In addition to discovering and advancing new technologies the **Science and Technology Office** would have a broader charter, having responsibility for maturing selected technologies, especially those identified in the Concept Formulation and Development Office.

The **Concept Formulation and Development Office** would have the charter to formulate, evaluate, and define concepts in each mission area and those that span multiple mission areas. The charter would include both novel systems concepts and novel operations concepts, and the organization would reflect this dual responsibility as shown in Figure 5.2. The Operational Concepts Office would be organized by mission area, while the Systems Concepts and Demonstration Office would be organized by broad system category. Each office would continually explore and evaluate new ways to integrate service capabilities and to make them more effective.

The mission area offices would accomplish the concept formulation task by encouraging consortia of operators and developers, including firms that now do not normally participate in defense markets, to propose new ways to gain military capability to meet the needs of the

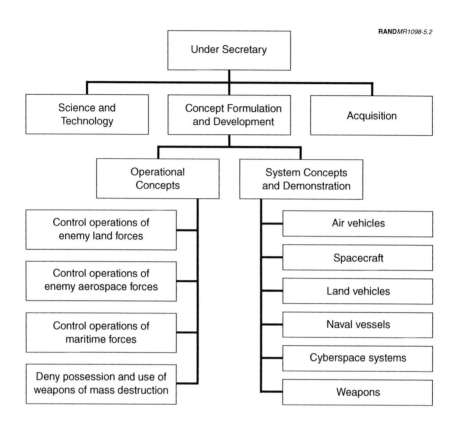

RAND*MR1098-5.2*

**Figure 5.2—Organization for the Operational Concepts Office and the
System Concepts and Demonstration Office**

future. Each office should encourage a robust competition among concepts and consortia aimed at generating more innovative, out-of-the-box approaches to gaining improved military capabilities. This kind of environment would be accomplished by convening a single or multiple concept option group(s) (COGs).[1]

---

[1]John Birkler, C. Richard Neu, and Glenn Kent, *Gaining New Military Capability: An Experiment in Concept Development*, Santa Monica, Calif.: RAND, MR-912-OSD, 1998.

Each COG would be an interactive partnership between those who know what is technically possible and those who know what is operationally viable and useful. Operational planners should lead the COG. It should include operators from the user commands, development planners from acquisition commands, scientists and engineers appropriate for each functional area in the operational concept, and a "red team" to identify possible countermeasures to the concepts being defined.

In the process of formulating new concepts, solutions would draw on existing and emerging technologies. For the latter case, the Under Secretary would task the Science and Technology (S&T) Office to plan and support S&T road maps to technological maturity. For the more technologically mature concepts, each office would conduct additional engineering-studies demonstrations to evaluate the technical feasibility, operational practicality, and robustness of the concepts. The demonstrations would also address doctrinal and command-control-communication issues raised by these concepts.

For selected concepts offering significant improvements in capability, the Operational Concepts Office would conduct the detailed end-to-end planning, estimate the numbers in more detail, and take into account problems of engineering and support—e.g., joint command and logistics support. Such planning would provide support for a proposal to milestone-D decision authority to proceed with concept demonstration.

Recognizing that many systems have high utility in multiple missions, we suggest that a second major office, the System Concepts and Demonstration Office, be established to demonstrate new ideas—post–milestone D. This group would explore the usefulness of a new design or concept in accomplishing a specific task or mission, or demonstrate a particular application of several integrated technologies. Also included in this office's mission are demonstrations of the system's ability to meet a specified threat and consideration of operations, support, and logistics. The System Concepts and Demonstration Office has the responsibility of integrating all the components and of providing an early and convincing demonstration of the operational concept. For each system demonstration project, a new project office would be formed to address that system concept and then disbanded when the task was completed.

After the successful demonstration and evaluation of a new concept, the Concept Formulation and Development Office would recommend one of the post-milestone-X acquisition options to the Under Secretary. For those concepts where further acquisition actions are authorized, the Under Secretary could in turn recommend to the Secretary of Defense which service should be assigned action responsibility.

The **Acquisition Office,** the third area of business, would then come in to play to oversee actions to acquire new systems and system elements approved by the Secretary of Defense.

PURPOSEFUL COMPETITION

A continuing weakness in the current system is that decisions on which service will provide forces that contribute capability toward a stated mission area or operational objective are made *ex ante.* That is, they are made before the service presents a set of options (concepts), and certainly before it has an opportunity to demonstrate how well the proposed concept might carry out the stated objectives. *Ex ante* allocation decisionmaking precludes the possibility of having many concepts to choose among.

A central element of our strategy is to stimulate competition of new ideas among the services. The best ideas would translate into that service being assigned the responsibility for implementing the concept and given the resources to perform that assignment. Competition managed in this manner would promote choices and thus set the stage to make informed choices among promising new concepts—choices made on the merit of the case, unhampered by a preconceived notion of "assignments" of particular missions and roles to a particular service. Of course, not every proposal should be funded. In fact, most probably should not be. The successful use of competition as a management tool requires the Secretary of Defense to make explicit decisions about winners and losers. Otherwise, the Department of Defense will proliferate solutions, squandering its resources on less competitive ideas.

And it is not just the services that should be included in the shaking-out process that competition provides. The areas currently most insulated from competition are those for which the defense agencies are responsible. The defense agencies were originally created to bring greater efficiency to the Department of Defense. Immediate savings could be had from a single overhead structure for each function, and it was hoped that a central decisionmaker would more rationally allocate the available resources, eliminate excess capacity, select the "best-practice" solution, and capitalize on economies of scale where they exist. While some defense agencies are certainly successful, the growth of defense agency resource levels relative to the DoD total is startling. Once created, there is little that others in DoD can do to challenge the agencies, for they occupy a monopoly position. Such a position in a declining budget environment virtually guarantees that their relative size will grow.

Chapter Six

IMPLEMENTING THE DUAL-PATH STRATEGY

The central goal of this study has been to propose a general strategy and associated process and organization that should enable military acquisition institutions to respond more quickly and effectively to service needs for novel weapon systems. We have described the proposed strategy, process, and organization in enough detail to show how they might achieve that goal, and we have outlined some of the major changes that would have to occur in the present institutions before the strategy, process, and organization could function effectively. But major changes in any large government bureaucracy are exceedingly hard to effect, and proposals for such change must be approached with respect for those difficulties. In this chapter, we explore those changes and how they might be implemented.

Many of the techniques suggested for moving systems along a path similar to the concept-demonstration path have been used before, and sometimes with spectacular success. Furthermore, most of the legislative provisions for such a process are already in place and have been used in ACTD and similar projects. But in every instance those programs were treated as exceptions to the "standard" process. Successfully implementing a concept-demonstration process will require that the concept be defined as part of the standard system.

We foresee three important obstacles to full implementation of the concept-demonstration path: (1) identifying a new system concept and justifying demonstration start, (2) creating an appropriate management attitude and perspective, and (3) providing funding. Each is discussed below.

IDENTIFYING AND JUSTIFYING A NEW DEMONSTRATION START

As noted in Chapter Three of this report, one of the key features needed to enable the accelerated processing of projects along the concept-demonstration path is to simplify and streamline the system for identifying candidate projects, defining the candidate system, and establishing the justification for starting demonstration. Today that process involves elaborate analysis and detailed documentation, neither of which is appropriate for novel systems involving important uncertainties and risks. An illustration of a desired process is drawn from a recent report on ACTD projects:[1]

> The Ships Capability Document (SCD) and CONOPS [concept of operations] were the primary documents that drove the design of the Arsenal Ship weapon system concept. The Navy's use of broad descriptions of desired performance, rather than specific requirements, was a major departure from the traditional acquisition approach. These two documents replaced the Mission Needs Statement (MNS), Operational Requirements Document (ORD), and Analysis of Alternatives (AOA) usually used.

> The SCD and CONOPS outlined the desired capabilities and suggested specific design attributes to provide them. Over a combined length of nine pages . . . the documents provided minimal performance specifications relative to system design. With few exceptions, the systems characteristics were defined as goals rather than hard requirements. Desired performance was described in broad terms; no specific method for achievement was suggested. . . . No formal systems specifications existed for the competing contractor teams; as a result, innovative design solutions were expected. . . . The Global Hawk Tier II+ HAE UAV is (another) attempt at radically reducing government control of weapon system's design. Two pages outlined the Global Hawk's mission description and preliminary concept of operations.

---

[1]Robert S. Leonard, Jeffrey A. Drezner, and Geoffrey Sommer, *The Arsenal Ship Acquisition Process Experience: Contrasting and Common Impressions from the Contractor Teams and Joint Program Office,* Santa Monica, Calif.: RAND, MR-1030-DARPA, 1999.

Experience with these and similar ACTD projects in the recent past shows that such an abbreviated process of program definition can be successful and is much faster than the standard process. We recommend that this process be adopted for projects intended for the concept-demonstration path, and that formal policy to that effect be issued.

CREATING A MANAGEMENT ORGANIZATION AND PROCESS

The notion of managing risks and uncertainties by building and testing a demonstration version of a novel system is contrary to present practice. To follow this path requires a mindset and outlook different from that of current MDAP management practices, including the notion that it is acceptable to fail occasionally.

We believe it will be necessary to establish a separate office to manage such projects, and that office should be in the office of the Under Secretary for Acquisition and Technology. Some key aspects of such an organization were outlined in Chapter Five of this report. We recommend that such an office be established, with authority to work with the services to foster development of novel systems. In addition to accumulating experience on how to best organize and manage such projects, the office would serve as an advocate for the initiation of new projects and be responsible for any special budget provisions made to support them. We believe that no new authorizing legislation would be required, but new legislation might be appropriate as part of the broader effort to secure congressional support for the general strategy.

FUNDING THE PROJECTS

One of the key risk-management processes built into the proposed concept-demonstration path is that, at the beginning of the program, no commitment is made beyond demonstration. Thus a minimum of funds are committed, and no serious force posture or employment strategies are put at risk. While this process appears sensible, it also runs counter to long-standing practices in the defense department to not allow projects to proceed past milestone I without substantial commitment to full development, production, and employment of the proposed system. That practice appears to have grown out of the difficulties encountered in canceling projects well after start and after a substantial constituency had been accumulated, together with the public criticism that often accompanied such cancellations (charges of waste, management incompetence, etc. have been encountered).

Even if the initial funding required by a concept-demonstration program is relatively small, such programs are not likely to survive in budgets of the individual services because of the limited funding available, and the intense pressure to use that funding to support traditional programs. Thus we recommend that a new budget category be established, and authorized through appropriate legislation to support at least the initial phases of concept-demonstration projects, up through the point where their capability would be tested in the field. The new management office in USD(A&T) noted above would be the agent responsible for requesting and managing such funds.

We believe that these three actions would provide the basic institutional foundation sufficient to enable a project-demonstration path to function as a standard element of the overall acquisition process.

FUNDING STRATEGIES FOR ACCELERATED ACQUISITION

One of the key strategies for accelerating the acquisition process, as advocated in the body of this report, involves starting development with a limited set of objectives so that lengthy concept refinement and program justification activities can be substantially truncated. One aspect of that strategy calls for committing funds at project start only sufficient to perform basic development and concept demonstration. Funding for any follow-on development and production would then be committed after successful demonstration of system costs and capabilities.

While that process should permit earlier start of many projects, especially those that entail important risks and uncertainties, a problem is encountered when transferring from concept demonstrator to full acquisition status. The Planning, Programming and Budgeting System (PPBS) process typically involves a planning cycle of two to four years' duration; projection of future funding for a new project must be injected into the planning cycle two to four years before funds are actually needed. Thus, if no funding action for full acquisition were taken until after demonstration of the system concept, a several-year gap would be introduced in the project. Some new procedures are needed for funding the later phases of such projects.

As part of this study, we examined two Army initiatives presently under way that were designed to assist in streamlining the Army's acquisition processes and reducing the time necessary to go from an idea/concept with "proof-of-concept" demonstrated to actual fielding of hardware.

The two programs are the Advanced Concepts and Technology II (ACT II) initiative and the Warfighting Rapid Acquisition Program (WRAP) initiative. The basic strategy employed in each initiative is that funds are appropriated by the Congress each year, in a "block-funding" approach, thus providing a "pool of resources" for the Army to use on selected projects for funding downstream without identification of individual projects to the Congress, at the time of their funding approval (one of the two WRAP projects does require subsequent notification and approval by the four defense committees prior to obligation of funds to a particular project).

ADVANCED CONCEPTS AND TECHNOLOGY II (ACT II) INITIATIVE

The ACT II initiative was started in FY94 through the collaborative efforts of the Army Chief of Staff and the Assistant Secretary of the Army for Research, Development and Acquisition (ASA(RDA)). The goal was to facilitate application of mature technology available in the commercial world and academia. Drawing on funds from R&D budget category 6.2 (exploratory development), this effort is focused on accomplishing technology demonstrations of a short duration and limited funding that would, if successful and evaluated critical to future Army needs by the Army user representative—Training and Doctrine Command (TRADOC), be selected for expedited development and ultimate production and fielding.[1] Initial funding for this initiative in FY94 was approximately $40 million but has now stabilized at approximately $12 million per year. Key characteristics of this initiative are that, for each project, funding is limited to $1.5 million (or less) and all efforts must be completed in 12 months (or less).

This initiative's process is managed by the Army Research Office-Washington (ARO-W), located and organizationally a part of HQ Army Materiel Command (AMC). As part of the Army's acquisition community, ARO-W is also responsible to ensure the unique ACT II acquisition/procurement processes are carried out, primarily through the AMC's major subordinate commands. However, all ACT II candidates are prioritized and selected for funding by TRADOC

---

[1]This information was obtained from a RAND meeting with Army Research Office-Washington (ARO-W) officials, Alexandria, Va., August 12, 1998.

and its Battle Labs. The materiel developer's key input is through technical evaluations and risk assessments of the proposals.

All ACT II candidates are proposed from industry or the academic community. All such efforts/proposals are only for hardware/staffing and do not include any studies/grants.

TRADOC and its Battle Labs play the key role *each year* in identifying focus areas and needs. These needs are then put into an ACT II Broad Agency Announcement (BAA) prepared and administered by ARO-W.[2] Initial concepts, suggested by industry/academia, are limited to two pages. From these submissions, concepts are evaluated and initially selected, and their industry nominees are then requested to prepare and submit technical and cost proposals of 25 pages or less.

These proposals are then evaluated by TRADOC and its Battle Labs with AMC technical support (through its Research Development and Engineering Centers). TRADOC then makes final selections of what is to be funded, and AMC acquisition organizations make the contractual awards. As stated earlier, these are one-year (or less) contracts, and not subject to renewal and second-year funding (as an ACT II effort). Successfully demonstrated projects, in the eyes of TRADOC, can then lead to these projects entering the acquisition process as WRAP candidates, ACT II projects, or entry into some other phase of acquisition.

The ACT II initiative began in FY94. There were some 63 technology demonstrations conducted in FY94 and FY95. Twenty-five projects were undertaken in FY96 and 20 in FY97.[3,4] A sample list of a few of the FY94–96 ACT II projects include

---

[2]Charter, Advanced Concepts and Technology II Initiative, signed by the Deputy Assistant Secretary of the Army for Research and Technology in the Office of the Assistant Secretary of the Army for Research, Development and Acquisition, Washington, D.C., revised April 1998.

[3]AMC/ARO-W ACT II Brochure, *Making Technology Work for Soldiers,* MG Roy Beauchamp, Deputy Chief of Staff for Research, Development and Acquisition in the Army Materiel Command (undated); obtained by RAND from ARO-W officials, August 12, 1998.

[4]*Jane's IDR EXTRA*, "Class Acts in the Making," Vol. 2, No. 8, August 1997.

- Tactical End-to-End Encryption Device
- GPS Drop Zone Assembly Aid
- Projectile Detection and Cueing System
- Test of a canard-controlled 155 mm projectile with roll-stabilized controls
- Integrated communications for the JTF commander
- Enhanced TPQ-37-theater missile defense point prediction.

A list of generic characteristics for these projects are provided below:

- Candidates nominated based on TRADOC operational needs
- All candidates from industry/academia
  — No studies/grants
- Due to limited scope/time for ACT II projects—mature "off-the-shelf" technology required such that a technology demonstration can be completed in 12 months or less.
- TRADOC is the ultimate selection and decisionmaker for ACT II projects
- Concept/proposal submissions limited in page count (2 pages for initial concept papers, 25 pages for proposals).

A few key observations can be made to assess the uniqueness of the ACT II initiative. First, it appears to be a common theme within the Army that the user—TRADOC, representing the ultimate user, the Commander in Chief of the Joint Task Force—is the focal point in the process and the key decisionmaker of what gets selected and funded. Army operational needs lead to topics selected for inclusion in the BAA. Also, a streamlined and abbreviated procurement process is followed by ARO-W and the procurement offices of AMC. Two-page concept papers and 25-page proposals are *all* that industry and academia are allowed to submit. Thus, industry is not obligated to spend considerable bid and proposal funds. These candidates must essentially be "off-the-shelf" mature technology, because the contract duration is 12 months or less and not renewable. There is essentially no time for extensive development efforts. Building, testing, and demonstration are all time allows. No follow-on efforts are promised; however, TRADOC, in evaluating demonstration results, can and does recommend highly successful projects (that would make a major positive contribution for warfighters) for continued acquisition efforts.

This initiative is part of the Army's S&T program and has little external visibility. It appears that no one from the ARO-W ACT II project management organization has met or briefed congressional staff members. All contact with Hill staffers is by officials at the Office of the Assistant Secretary of the Army for Research, Development and Acquisition. Yet the ACT II initiative appears stable and is being accomplished through cooperation and support of the Office of ASA(RDA), the Office of the Deputy Chief of Staff for Operations (HQ Army), AMC, and TRADOC leadership.

WARFIGHTING RAPID ACQUISITION PROGRAM (WRAP) INITIATIVE

The WRAP initiative was created by the combined leadership of the Army Chief of Staff and the Commanding General (CG) of the U.S. Army Training and Doctrine Command. They realized a need to fund critically needed systems coming out of the Advanced Warfighting Experiments (AWEs) that add significant operational utility for the soldier operating in the Army's Force XXI environment without waiting for the two-year PPBS process to initiate funding.[5]

WRAP essentially started with Army 4-star discussions on the Hill in the spring of 1996 where agreement was reached to provide funds (6.4 RDT&E dollars) for accelerating development and fielding of selected hardware/software. Congressional committees offered to include $50 million for FY97 if the Army agreed to budget $100 million per year beginning in FY98. The Army agreed. A key feature of this initiative was that the funds would be appropriated by Congress without the prior identification of what specific projects were to be undertaken. However, agreement was made that prior to the Army obligating any funds for a specific project, the Army would identify the projects to the four defense committees on the Hill and obtain their written approval. In addition, these WRAP projects were to be

---

[5]*Army RD&A Magazine*, January–February 1998, pp. 10–11.

considered part of the acquisition process and not something outside of the acquisition process; the controlling document is the Department of the Army Policy Memo cosigned by the Vice Chief of Staff of the Army (VCSA) and the ASA(RDA) of April 1996.[6]

To be approved as a WRAP project, a candidate must be nominated by the CG TRADOC and proceed through an approval process, ending up at a WRAP Army System Acquisition Review Council (ASARC) meeting cochaired by the Military Deputy to the ASA(RDA) and the Assistant Deputy Chief of Staff for Operations and Plans, Force Development. Once approved, these programs are managed within the Army's acquisition organization under the leadership of the ASA(RDA). Some common characteristics of these programs are listed below:

- Based on results of AWEs and other demonstrations/simulations
- Part of acquisition process/decision authority
  - ASARC approval required
  - Hill notification/approval required
- All WRAP projects are small programs (ACAT III/IV)
  - New initiative—not to pay old bills
  - Not for indefinite experimentation
  - Must be within two years of milestone III
  - Key for first digitized division/Force XXI
  - Must have production dollars approved through POM
  - Must have initial operational test and evaluation conducted
- An operational requirement document (ORD) not required until milestone III (TRADOC statement of need at WRAP initiation).

Most of the projects undertaken are in the $5–$15 million range (for two years' funding) to accomplish necessary EMD efforts. Examples

---

[6]Memorandum, "Policy for Warfighting Rapid Acquisition Program," cosigned by the VCSA and the ASA(RDA), April 11, 1996.

include such projects as (see *Army RD&A Magazine*, January–February 1998, p. 12):

- Mortar Fire Control System

- Army Airborne Command and Control System

- Striker (HUMV with LD/other sensors)

- Radio Frequency Tags

- Tactical Internet

- Combat Synthetic Trainer Assessment Range

- Lightweight Laser Designator Rangefinder

- Gun Laying Position System.

For FY99 project selections, the ASARC approval process was moved forward to allow an early November 1998 ASARC to facilitate congressional notification and approval, and thus permit earlier program execution in the fiscal year.

The WRAP ASARC follows from a structured process[7] of approximately six months' duration. TRADOC's call for candidates for the FY99 selection occurred in May 1998, with initial ideas and topics submitted in June 1998. Documentation for the candidates was due in a July/August time frame, with initial TRADOC recommendations submitted to HQ, Department of the Army (DA), in September 1998. DA staff reviews took place in September 1998, with feedback to TRADOC; the TRADOC WRAP Battle Lab Board of Directors reviewed and prioritized final candidates in October, which then led to the scheduled November 1998 WRAP ASARC.

Some interesting observations regarding the WRAP initiative have been gained from discussions with senior Army staff personnel.[8] They attribute program success to strong 4-star sponsorship and involvement from both the Army Chief of Staff and the CG TRADOC.

---

[7]Data obtained by RAND from the Office of the ASA(RDA) briefing to House Appropriations staff on status of the WRAP initiative, August 10, 1998.

[8]RAND meeting with Office of the ASA(RDA) personnel, *Wrap Initiatives Status and Process*, Arlington, Va., August 11, 1998.

WRAP implementation has had challenges and problems and needed continuous interaction with congressional staff members. In fact, interactions with staffers on specific projects to be undertaken have been necessary and frequent.

FY98 funding was approved by Congress at approximately $100 million but was subsequently reduced by $27 million by the Army leadership. The WRAP FY99 funding request was reduced by the Hill to approximately $65–$70 million (the exact amount will be ironed out in a conference committee). The chief cause of the reduction was the end result of the Army action with FY98 funds (reduced from $100 million to $73 million).

COMMON THEMES AND OBSERVATIONS

There are four features that stand out as key to the success of the ACT II and WRAP initiatives. These features do not, in and of themselves, guarantee success but at least indicate to those within the Department of the Army, OSD, and the Congress that ACT II and WRAP are important for Army warfighters. The four are listed below:

- Strong Army (uniformed) support and involvement at the top.

- "User" control of project process/selection of candidates, based on horizontally integrated priority operational needs.

- Single structured and documented candidate review process (for each program) that involves all levels of the Army's TRADOC community and accepted as fair and objective.

- Block funding for each initiative at an affordable level (not excessive) that is accepted at the DA level, the OSD level, and by the Congress.

The first item is key for any acquisition initiative. Strong top-level leadership must be involved. The Army Chief of Staff and CG TRADOC involvement in setting up the process for user control in developing priority needs and in having essentially final say in which candidates are selected for funding in each initiative indicates that the user is really "in charge" of setting priorities. Over the past couple of decades and particularly in the past few years, the Army has developed the approach of expanding user participation in resource allocation. The establishment of Battle Labs at Army functional centers to help develop strategy, doctrine, and tactics in the modern and technologically fast-paced environment, and to assist in establishing a horizontal linkage across these functional boundaries has assisted in establishing credibility to the outside world that the Army knows what it needs.

It is conceivable that the Army's approach to and process of user involvement could be expanded to operate in the joint arena, with the JCS and USD(A&T) being the senior DoD leadership fostering a joint approach for DoD initiatives, similar to that which exists in the Army. However, with continued shrinking of DoD resources, bold changes in the current processes must occur to maximize the benefit of available resources.